β Beta

Multiple-Digit Addition and Subtraction

Tests

Math·U·See

1-888-854-MATH (6284) - mathusee.com
sales@mathusee.com

Beta Tests: Multiple-Digit Addition and Subtraction
©2012 Math-U-See, Inc.
Published and distributed by Demme Learning

mathusee.com

1-888-854-6284 or +1 717-283-1448 | demmelearning.com
Lancaster, Pennsylvania USA

ISBN 978-1-60826-073-7
Revision Code 1218-G

Printed in the United States of America by The P.A. Hutchison Company
4 5 6 7 8 9 10

For information regarding CPSIA on this printed material call: 1-888-854-6284
and provide reference #1218-03022022

1

Count and write the number. Then say it.

1.

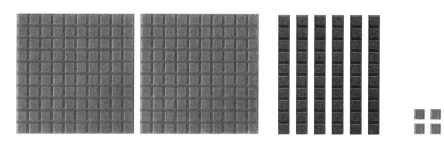

_____   _____ _____

2.

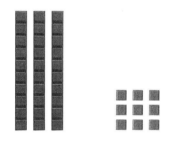

_____   _____

Build and say the number.

3. 421   4. 207

Add.

5.   0
  + 7
  _____

6.   8
  + 3
  _____

Add.

7.　　9
　　+ 7

8.　　6
　　+ 5

9.　　4
　　+ 4

10.　　3
　　+ 6

11.　　8
　　+ 2

12.　　4
　　+ 7

13. 5 + 5 = _____

14. 3 + 5 = _____

15. 6 + 4 = _____

16. 5 + 7 = _____

17. 9 + 1 = _____

18. 9 + 6 = _____

Put the numbers in order from the least to the greatest.

1. 16, 34, 20

   _____ , _____ , _____

2. 99, 19, 49

   _____ , _____ , _____

Put the numbers in order from the greatest to the least.

3. 108, 82, 83

   _____ , _____ , _____

4. 11, 111, 101

   _____ , _____ , _____

Fill in the blanks with the correct number for the sequence.

5. _____ , 23, _____ , _____ , 26

Build and say the number.

6. 140

7. 22

Add.

8. $\begin{array}{r} 4 \\ + \ 0 \\ \hline \end{array}$

9. $\begin{array}{r} 2 \\ + \ 7 \\ \hline \end{array}$

10. $\begin{array}{r} 5 \\ + \ 9 \\ \hline \end{array}$

11. $\begin{array}{r} 3 \\ + \ 6 \\ \hline \end{array}$

12. 2 + 3 = _____

13. 7 + 7 = _____

14. 9 + 1 = _____

15. Chad made 14 points. Craig made 24 points. Whose number of points is less?

_____

16. We had eight arrowheads in our collection. While on vacation we found nine more. How many arrowheads do we have now?

_____

3

Compare. Then fill in the oval with <, >, or =.

1. $2 + 9$ ◯ $6 + 1$ 2. $4 + 5$ ◯ $5 + 4$

3. 19 ◯ 91 4. 111 ◯ 101

5. $7 + 6$ ◯ $9 + 4$ 6. $6 + 3$ ◯ $5 + 5$

Fill in the blanks with the correct number for the sequence.

7. _____, 28, _____, _____, 25

Add.

8. $\begin{array}{r} 2 \\ + 8 \\ \hline \end{array}$ 9. $\begin{array}{r} 6 \\ + 9 \\ \hline \end{array}$

Add.

10.   3
     + 4
    _____

11.   0
     + 2
    _____

12.   7 + 8 = _____

13.   4 + 7 = _____

14.   9 + 4 = _____

15.   Rhonda compared 9 + 5 and 6 + 6. Write an inequality showing which sum is less.

    _____

16.   Coral opened her birthday card and found five dollars. In the next card there was no money. How many dollars did Coral have altogether?

    _____

Round to the nearest ten.

1. 85 → _____   2. 33 → _____

3. 52 → _____

Round to the nearest ten and estimate the answer.

4.   5 2   (   )   5.   4 1   (   )
     + 1 7 (   )   + 2 5 (   )
    ———— (   )   ———— (   )

6.   3 2   (   )   7.   7 7   (   )
     + 2 5 (   )   + 1 2 (   )
    ———— (   )   ———— (   )

Add across and down. Then add your answers and see if they match.

8.

4	4	
3	2	

9.

1	6	
5	3	

Compare. Then fill in the oval with <, >, or =.

10. 5 + 5 $\bigcirc$ 6 + 1

11. 399 $\bigcirc$ 329

12. I already ate two muffins. Mom gave me two more. If I eat them also, how many muffins will I have eaten in all?

_____

13. A grocer sold 27 cans of olives last week and 31 cans this week. Estimate how many cans of olives the grocer sold the last two weeks.

_____

14. Write an inequality showing which is greater: 21 or 201.

_____

Write using place-value notation.

1. 628 = _____ + _____ + _____

2. 49 = _____ + _____

Add using place-value notation.

3.   1 6   →   1 0 + 6
    + 3 2   →   3 0 + 2

4.   3 8 0   →   3 0 0 + 8 0 + 0
    + 5 1 6   →   5 0 0 + 1 0 + 6

5.   5 4   →   5 0 + 4
    + 2 2   →   2 0 + 2

6.   4 3 3   →   4 0 0 + 3 0 + 3
    + 4 2 5   →   4 0 0 + 2 0 + 5

Round to the nearest ten and estimate. Then find the exact answer.

7.
$$
\begin{array}{r}
3\ 4 \\
+\ 6\ 1 \\
\hline
\end{array}
$$
( )
( )
( )

8.
$$
\begin{array}{r}
1\ 2 \\
+\ 1\ 6 \\
\hline
\end{array}
$$
( )
( )
( )

Compare. Then fill in the oval with <, >, or =.

9. $4 + 6 \bigcirc 6 + 4$

10. $49 \bigcirc 94$

Solve for the unknown.

11. $7 + X = 14$

12. $5 + Y = 8$

13. $8 + A = 16$

14. I saw 45 deer and 12 moose in the park. How many large animals did I see?

_____

15. Riley saved 132 pennies and 160 nickels. How many coins has she saved?

_____

Skip count by two and write the numbers.

1. _____, _____, 6, _____, _____, _____, _____, _____, _____, _____

Skip count by two to find the answer.

2. How many ice cream cones? _____

Add using place-value notation.

3. $\begin{array}{l} 5\,5 \rightarrow 5\,0 + 5 \\ +\ \ 2 \rightarrow \underline{0\,0 + 2} \end{array}$

4. $\begin{array}{l} 1\,0\,6 \rightarrow 1\,0\,0 + 0\,0 + 6 \\ +2\,2\,1 \rightarrow \underline{2\,0\,0 + 2\,0 + 1} \end{array}$

Round to the nearest ten and estimate. Then find the exact answer.

5. $\begin{array}{r} 4\ 4 \\ +\ 1\ 3 \\ \hline \end{array}$ ( )
( )
( )

6. $\begin{array}{r} 3\ 8 \\ +\ 2\ 1 \\ \hline \end{array}$ ( )
( )
( )

Solve for the unknown.

7. $2 + F = 8$

8. $5 + G = 13$

9. $9 + H = 18$

10. Justin had five toy trucks. His dad gave him nine more. How many trucks does he have now?

_____

11. Three people raised both hands in the air. Skip count to find the number of hands.

_____

12. Bekka has 40 red jelly beans and 25 green jelly beans. How many does she have in all?

_____

Add. Regroup if needed.

1.
$$\begin{array}{r} 3\ 5 \\ +\ 2\ 9 \\ \hline \end{array}$$

2.
$$\begin{array}{r} 1\ 6 \\ +\ 6\ 7 \\ \hline \end{array}$$

3.
$$\begin{array}{r} 2\ 9 \\ +\ 4\ 4 \\ \hline \end{array}$$

4.
$$\begin{array}{r} 1\ 9 \\ +\ 4\ 6 \\ \hline \end{array}$$

5.
$$\begin{array}{r} 5\ 4 \\ +\ \ \ 7 \\ \hline \end{array}$$

6.
$$\begin{array}{r} 7\ 5 \\ +\ 2\ 2 \\ \hline \end{array}$$

Skip count by two and write the numbers.

7. ____, ____, ____, ____, ____, ____, ____, ____, ____, 20

Compare. Then fill in the oval with <, >, or =.

8. 7 + 8 $\bigcirc$ 9 + 4

9. 21 $\bigcirc$ 12

10. 801 $\bigcirc$ 810

Solve for the unknown.

11. $4 + X = 5$

12. $9 + W = 12$

13. $4 + A = 7$

14. Douglas grows strawberries to sell. Last year he sold 55 quarts. This year he sold 39 quarts. How many quarts of strawberries did he sell in all?

_____

15. Before nightfall, 212 zebras stopped to drink at the water hole. They were joined by 344 antelope. How many animals were at the water hole?

_____

Count and write the number. Then say it.

1.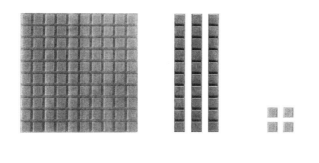

   _____   _____   _____

Put the numbers in order from the least to the greatest.

2. 111, 17, 71   _____, _____, _____

Compare. Then fill in the oval with <, >, or =.

3. 2 + 9 ◯ 7 + 3   4. 8 + 6 ◯ 6 + 8

5. 342 ◯ 324

Round to the nearest ten.

6. 18 → _____   7. 35 → _____

8. 79 → _____

Solve for the unknown.

9. 5 + X = 7   10. 8 + Q = 16

11. 5 + B = 9

Skip count by two and write the numbers.

12. ____, ____, ____, ____, ____, ____, ____, ____, 18, ____

Add. Regroup if needed.

13. $\begin{array}{r} 1\ 6 \\ +\ 2\ 1 \\ \hline \end{array}$

14. $\begin{array}{r} 5\ 5 \\ +\ 3\ 7 \\ \hline \end{array}$

15. $\begin{array}{r} 1\ 5 \\ +\ 2\ 5 \\ \hline \end{array}$

16. $\begin{array}{r} 7\ 8 \\ +\ \ \ 8 \\ \hline \end{array}$

17. $\begin{array}{r} 2\ 3\ 4 \\ +\ 1\ 4\ 2 \\ \hline \end{array}$

18. $\begin{array}{r} 3\ 6\ 1 \\ +\ 2\ 0\ 5 \\ \hline \end{array}$

19. Sally wrote two letters every day. Skip count to find how many letters she wrote in four days.

_____

20. Penny counted 28 green cars and 43 red cars on the trip to Grandma's house. Estimate how many cars she counted in all. Then find the exact answer.

_____

21. A fisherman caught 415 big fish and 221 little fish. How many fish did he catch in all?

_____

22. Alexis bought a dress for 39 dollars and a hat for 15 dollars. How much money did she spend altogether?

_____

Skip count by ten and two. Write the numbers.

1. _____, _____, _____, 40, _____, _____, _____, _____, _____, _____

2. _____, 4, _____, 8, _____, _____, _____, _____, _____, _____

Add. Regroup if needed.

3.　　4 5
　　+ 2 7

4.　　6 5
　　+ 1 6

5.　　3 7
　　+ 3 4

Add. These do not need regrouping.

6.　　1 0 9
　　+ 2 9 0

7.　　4 7 7
　　+ 1 2 2

8.　　8 3 4
　　+ 　4 5

Solve for the unknown.

9. 4 + X = 8

10. 5 + Y = 14

11. 7 + R = 10

12. Everyone in our family has ten toes. There are seven people in the family. Skip count to find how many toes in all.

    _____

13. Noah had four dimes in his pocket. Then he found one more dime. How many cents does Noah have now?

    _____

14. René wrote six stories about horses, one story about birds, and three stories about sailboats. How many stories did she write altogether?

    _____

15. Thirty-four boys and forty-seven girls came to the picnic. How many children came to the picnic?

    _____

Skip count by five and ten. Write the numbers.

1. _____, _____, 15, _____, _____, _____, _____, _____, _____, _____

2. _____, _____, 30, _____, _____, _____, _____, _____, _____, _____

Add. Regroup if needed.

3.   2 6
   + 5 5

4.   6 8
   +   8

5.   1 3
   + 4 9

Add. These do not need regrouping.

6.   1 1 6
   + 4 3 1

7.   6 9 1
   + 3 0 5

8.   5 0 0
   + 2 1 6

Solve for the unknown.

9. $7 + R = 13$

10. $5 + A = 9$

11. $5 + X = 5$

Draw lines to match the questions with the right answers.

12. What coin is worth one cent?

13. What coin is worth five cents?

14. What coin is worth ten cents?

15. Caitlyn's mom gave her six boxes for her rock collection. Caitlyn put five rocks in each box. Skip count to find how many rocks she has.

_____

Fill in the blanks and say the amount.

1.

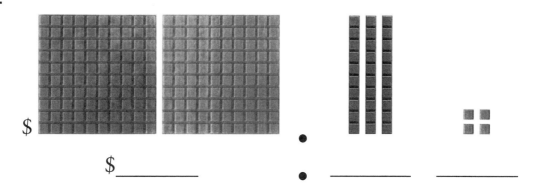

$

$_____ • _____ _____

Build and say.

2. $1.28

3. $2.06

4. $3.41

5. $4.15

Skip count to find how the number of cents.

6.

= _____ ¢

Add.

7.
```
  6 3
+ 1 7
```

8.
```
  2 3 4
+ 1 4 5
```

9.
```
  4 8
+ 2 6
```

Round to the nearest tens place.

10. 45 → _____

11. 18 → _____

12. 32 → _____

13. Sheri earned nine dollars, five dimes, and four pennies. How much money did she earn?

_____

14. Pearl made trail mix. She used three pounds of raisins, two pounds of cashews, and five pounds of peanuts. How many pounds of trail mix did she make?

_____

15. Angela has two dimes and five nickels. How much money does she have?

_____

Round to the nearest hundred.

1. 126 → _____   2. 314 → _____

3. 509 → _____

Add. Regroup if needed.

4.   1 3 2
  + 4 1 8

5.   6 7 9
  + 2 7 6

6.   5 2 0
  + 1 8 8

7.   8 7
  + 2 8

8.   4 5
  + 4 2

9.   3 9
  + 1 5

Review your subtraction facts.

10.   7
  – 2

11.   5
  – 1

Review your subtraction facts.

12.   8
     – 0

13.   1 1
     – 9

14.   5
     – 3

15.   1 0
     – 2

16.   6
     – 1

17.   7
     – 5

Skip count by five and write the numbers.

18. ____, ____, ____, ____, 25, ____, ____, ____, ____, ____

19. Cameron had 476 stamps in his collection at the beginning of the year. Since then he has collected 125 more. How many stamps does he have now? Estimate first and then solve.

_____

20. Sara is six years old. In how many years will she be ten?

_____

Add. Regroup if needed.

1.　$2.4 6
　　+ 1.7 9

2.　$3.9 1
　　+ 3.2 5

3.　$1.5 7
　　+ 0.3 8

4.　　4 0 1
　　+ 5 7 9

5.　　6 5
　　+ 1 1

6.　　2 2
　　+ 　8

Subtract.

7.　1 8
　　− 9

8.　　8
　　− 4

9.　1 3
　　− 8

10.　1 6
　　− 9

11.　1 0
　　－ 5

12.　1 2
　　－ 8

13.　1 4
　　－ 7

14.　1 3
　　－ 9

Fill in the blanks and say the amount.

15.

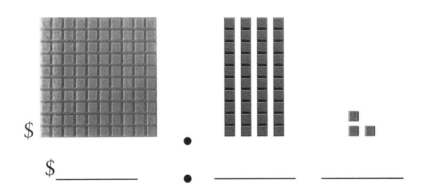

$ _____

$_____　•　_____　_____

16. A hamburger costs $1.89, and a shake costs $2.36. What is the cost of the meal?

_____

17. Duncan has four dimes and five nickels. Use a dollar sign and decimal point to write how much money he has in all.

_____

18. Fifteen people entered the race, but only nine finished it. How many people did not finish the race?

_____

Add the columns. Find 10 first if you can.

1.
```
    6
    4
 +  7
_____
```

2.
```
    3
    2
    7
 +  8
_____
```

3.
```
   1 1
   4 2
   3 9
 + 3 6
_____
```

4.
```
   $1.2 6
 +  8.4 7
_____
```

5.
```
   6 4 1
 +   9 0
_____
```

6.
```
   5 8
 +   2
_____
```

Subtract.

7.
```
   1 0
 -   7
_____
```

8.
```
    9
 -  3
_____
```

9.
```
   1 0
 -   6
_____
```

10.
```
    9
 -  4
_____
```

11. 1 0
     − 5

12.   9
     − 7

13. 1 0
     − 2

14. 1 0
     − 3

Skip count and write the numbers.

15. _____, 4, _____, 8, _____, _____, _____, _____, _____, _____

Match the shapes with their names.

16.   square

17.   rectangle

18.   triangle

19. Brooke rode her bike for three hours on Monday, four hours on Tuesday, and six hours on Wednesday. On Thursday she was tired and rode only one hour. How many hours has she ridden in all?

    _____

20. Ian has 6 nickels, 3 dimes, and 15 pennies. How much money does he have?

    _____

Measure using a ruler.

1.  _____ "

2. _____ "

Add.

3. $1.3 4
    + 2.0 7

4.   8 6 6
    +   3 6

5.   2 6
    + 4 4

6. 1 + 1 + 3 + 9 + 7 = ____

7. 13 + 4 + 11 + 6 = ____

Compare. Then fill in the oval with <, >, or =.

8. 17 – 8 ◯ 15 – 7

Subtract.

9. 13 – 7 = _____

10. 15 – 9 = _____

11. 7 – 3 = _____

12. 14 – 6 = _____

13. 7 – 4 = _____

14. 12 – 7 = _____

15. How many sides does a triangle have?

_____

16. How many sides does a rectangle have?

_____

17. Jill had eleven chores to do. She did four of them. Her sister helped by doing two more. How many chores does Jill have left to do?

_____

18. Jody is four feet tall. How many inches tall is she? (Use column addition.)

_____

Circle the name of the shape. Add to find the perimeter.

1. The shape is a:

square   rectangle   triangle

The perimeter is _____ inches.

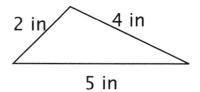

2. The shape is a:

square   rectangle   triangle

The perimeter is _____ inches.

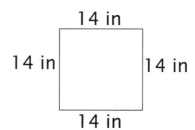

Add.

3. $2.2 2
     + 3.6 8

4.   7 7 1
     + 1 3 5

5.   4 6
    + 5 5

Subtract.

6. $13 - 9 =$ _____

7. $11 - 4 =$ _____

8. $13 - 4 =$ _____

9. $12 - 5 =$ _____

10. $9 - 6 =$ _____

11. $11 - 3 =$ _____

12. Lydia has a garden shaped like a rectangle. Two sides are five feet long. The other two sides are six feet long. What is the perimeter of her garden?

_____

13. A yardstick is three feet long. Use column addition to find the number of inches in a yard.

_____

14. Alex found three dimes and three nickels. He spent 5¢ for gum. How much money does he have left?

_____

15. Mom baked nine pies for the family reunion. The dog ate three pies. How many pies are left?

_____

Compare. Then fill in the oval with <, >, or =.

1. 12 - 6 $\bigcirc$ 3 + 3   2. 14 - 9 $\bigcirc$ 15 - 8

3. 117 $\bigcirc$ 171

Round to the nearest ten.

4. 47 → ____   5. 92 → ____

6. 65 → ____

Round to the nearest hundred.

7. 108 → ____   8. 355 → ____

9. 729 → ____

Skip count and write the numbers.

10. ____, 4, ____, ____, 10, ____, ____, ____, ____, ____

11. ____, ____, 15, ____, ____, ____, ____, ____, 45, ____

12. ____, 20, ____, 40, ____, ____, ____, ____, ____, ____

Add.

13.
```
   1 3
 + 4 5
```

14.
```
   6 4
 + 2 8
```

15.
```
   7 6
 + 1 9
```

16.
```
  $3.6 1
 +  1.7 9
```

17.
```
   4 5 2
 + 2 5 6
```

18.
```
   9 6 8
 +   7 5
```

Add the columns. Find 10 first if you can.

19.
```
   7
   3
 + 2
```

20.
```
   5
   4
   5
 + 1
```

21.
```
   6 5
   1 2
   1 8
 +   3
```

Circle the name of the shape. Add to find the perimeter.

22. The shape is a:

    square   rectangle   triangle

    The perimeter is _____ inches.

21 in

19 in   19 in

21 in

Subtract.

23.   1 4
      – 5
      ———

24.   1 8
      – 9
      ———

25.   1 0
      – 4
      ———

26.   1 7
      – 8
      ———

27.   7
      – 5
      ———

28.   1 1
      – 6
      ———

29. Morgan bought four feet of ribbon. How many inches of ribbon does she have?

    _____

30. Naomi counted her money. She has three dimes, six nickels and seven pennies. How much money does she have?

_____

Write the number and say it.

1. $8,000 + 600 + 70 + 2 =$ _____

2. $90,000 + 3,000 + 100 + 40 + 5 =$ _____

3. Two hundred thirty-six thousand, one hundred seventy-nine = _____

4. Eleven thousand, four hundred sixteen = _____

Add.

5.
```
  7 6 3
+ 5 1 8
```

6.
```
  3 5 4
+ 9 5 6
```

7.
```
$3.5 6
+ 2.4 9
```

8.
```
  5 6
  4 4
+ 9 8
```

Subtract.

9. $14 - 9 =$ _____

10. $13 - 5 =$ _____

11. 16 − 7 = _____   12. 11 − 4 = _____

13. 10 − 6 = _____   14. 12 − 8 = _____

15. Which is more money, four nickels or three dimes?

_____

16. Clyde made four birthday cards on Monday and seven cards on Tuesday. On Wednesday he sent eight cards to his friends. How many cards does he have left?

_____

Round to the nearest thousand.

1. 2,415 → _____

2. 9,503 → _____

Add.

3.
```
  6, 3 0 9
+ 1, 7 1 2
```

4.
```
  8, 4 1 6
+ 3, 5 5 4
```

5.
```
  $8.9 2
+  4.2 5
```

Write the number and say it.

6. Seven hundred eighty-six thousand, four hundred ten

_____

Add to find the perimeter.

7.

12 ft, 20 ft, 16 ft _____

Subtract.

8. $10 - 3 =$ _____

9. $9 - 5 =$ _____

10. $15 - 6 =$ _____

11. $12 - 4 =$ _____

12. $14 - 7 =$ _____

13. $7 - 4 =$ _____

14. Fourteen birds landed on my lawn. Five of them were blue jays. The rest were robins. How many were robins?

_____

15. Dennis counted 2,176 ants going down into the ground. Danny counted 3,402 ants. How many ants did the boys count altogether?

_____

16. Thomas spent $29 for a book, $21 for a game, and $9 for lunch. How much did he spend in all?

_____

Add.

1.
```
  1 2 3
  6 7 8
  2 0 7
  1 3 3
+ 3 1 2
```

2.
```
  6 6 2
  1 0 8
  5 0 0
  5 4 3
+ 1 6 1
```

3.
```
  7 8 2
  1 5 3
  2 6 9
  3 4 1
+ 8 0 7
```

Write the number and say it.

4. 800,000 + 70,000 + 1,000 + 400 + 60 + 5 _____

5. Fifty-six thousand, two hundred seventeen _____

Subtract.

6.
```
  1 5
-   9
```

7.
```
  1 0
-   8
```

8.
$$
\begin{array}{r}
1\ 3 \\
-\ 9 \\
\hline
\end{array}
$$

9.
$$
\begin{array}{r}
5 \\
-\ 3 \\
\hline
\end{array}
$$

10.
$$
\begin{array}{r}
1\ 7 \\
-\ 8 \\
\hline
\end{array}
$$

11.
$$
\begin{array}{r}
1\ 8 \\
-\ 9 \\
\hline
\end{array}
$$

12.
$$
\begin{array}{r}
7 \\
-\ 6 \\
\hline
\end{array}
$$

13.
$$
\begin{array}{r}
1\ 1 \\
-\ 8 \\
\hline
\end{array}
$$

Skip count and write the numbers.

14. 2, _____, _____, _____, _____, _____, _____, _____, 18, _____

15. Ada was training for the big race. She walked the following numbers of miles: 5, 7, 4, 10, and 13. How many miles did she walk in all?

_____

16. Fifteen people wanted to buy the new game, but only seven of the games were left at the store. How many people were disappointed?

_____

Add.

1.
$$\begin{array}{r} 2\,8\,3\,4 \\ 1\,5\,4\,8 \\ +\ 3\,6\,7\,2 \\ \hline \end{array}$$

2.
$$\begin{array}{r} 4\,5\,0\,6 \\ 3\,2\,9\,4 \\ +\ 2\,7\,5\,3 \\ \hline \end{array}$$

3.
$$\begin{array}{r} 6\,7\,2\,9 \\ 5\,3\,2\,6 \\ +\ 8\,3\,6\,1 \\ \hline \end{array}$$

Subtract.

4.
$$\begin{array}{r} 1\,0 \\ -\ 6 \\ \hline \end{array}$$

5.
$$\begin{array}{r} 1\,6 \\ -\ 9 \\ \hline \end{array}$$

6.
$$\begin{array}{r} 6 \\ -\ 3 \\ \hline \end{array}$$

7.
$$\begin{array}{r} 1\,1 \\ -\ 4 \\ \hline \end{array}$$

8.
$$\begin{array}{r} 9 \\ -\ 5 \\ \hline \end{array}$$

9.
$$\begin{array}{r} 1\,4 \\ -\ 7 \\ \hline \end{array}$$

10. $\begin{array}{r} 1\ 1 \\ -\ 6 \\ \hline \end{array}$ 11. $\begin{array}{r} 1\ 0 \\ -\ 5 \\ \hline \end{array}$

Compare. Then fill in the oval with <, >, or =.

12. 5 – 4 ◯ 9 – 8   13. 13 + 5 ◯ 13 – 5

14. 209 ◯ 902

15. A scientist studied 1,367 oak trees, 2,079 maple trees, and 1,534 pine trees. How many trees did he study altogether?

_____

16. Donald drew a triangle. Each side was 10 inches long. What was the perimeter of the triangle?

_____

17. Seven people splashed their feet in our pool. How many toes are in the pool? (skip count)

_____

18. Grace had 46¢. She found one nickel. How much money does she have now?

_____

Subtract and check by adding.

1.
```
   5 0
 - 2 0
```

2.
```
   4 2
 - 3 0
```

3.
```
   8 5
 - 2 3
```

4.
```
   2 6 3
 - 1 1 2
```

5.
```
   8 8 8
 - 5 4 6
```

6.
```
   3 7 4
 -   6 2
```

Add.

7.
```
   3 2 7 4
   1 6 9 0
 + 7 2 1 6
```

8.
```
   5 2 0 9
   6 1 2 8
 + 1 3 4 2
```

9.
```
   1 2 3
   4 5 6
   7 8 9
 + 1 1 1
```

Write the number and say it.

10. 50,000 + 4,000 + 900 + 70 + 1 = _____

Skip count and write the numbers.

11. ____, 4, ____, ____, ____, ____, ____, ____, ____, ____

12. ____, 10, ____, ____, ____, ____, ____, ____, ____, ____

13. ____, 20, ____, ____, ____, ____, ____, ____, ____, ____

14. Richard is 29 years old, and Joanne is 26 years old. How many years older is Richard?

_____

15. Mom made 48 cookies for the party. The guests ate 34 cookies. How many cookies are left?

_____

Write the number of minutes shown by each clock.

1.

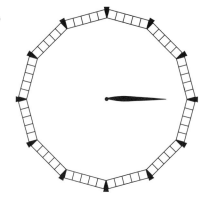

_____

2.

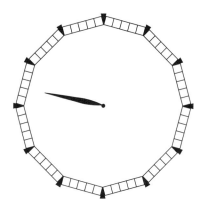

_____

3.

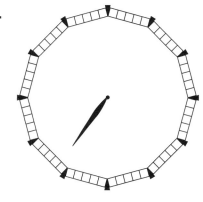

_____

4.

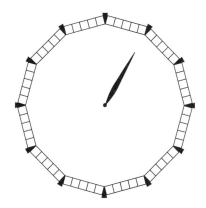

_____

Subtract and check.

5.   7 8
    −   6

6.   5 6 2
    −   4 1

7.   8 2 6
    − 6 0 3

Add.

8. $7.2 0
    + 4.5 6

9.   6 0 3 4
    + 2 9 8 7

10.   5 9 4
       3 4 6
    + 2 7 3

11. Kayla will be 25 in 13 years. How old is she now?

    _____

12. Neil drove 38 miles on Monday, 107 miles on Tuesday, and 73 miles on Wednesday. How many miles did he drive in all?

    _____

Subtract, using regrouping when needed. Check by adding.

1.
$$\begin{array}{r} 6\ 2 \\ -\ 2\ 8 \\ \hline \end{array}$$

2.
$$\begin{array}{r} 3\ 4 \\ -\ 1\ 5 \\ \hline \end{array}$$

3.
$$\begin{array}{r} 9\ 3 \\ -\ 5\ 6 \\ \hline \end{array}$$

4.
$$\begin{array}{r} 2\ 2 \\ -\ 1\ 8 \\ \hline \end{array}$$

5.
$$\begin{array}{r} 8\ 1\ 7 \\ -\ 2\ 0\ 5 \\ \hline \end{array}$$

6.
$$\begin{array}{r} 6\ 2\ 3 \\ -\ 5\ 1\ 2 \\ \hline \end{array}$$

Add.

7.
$$\begin{array}{r} \$4.3\ 5 \\ +\ 2.9\ 8 \\ \hline \end{array}$$

8.
$$\begin{array}{r} 4\ 8\ 2\ 6 \\ 1\ 4\ 9\ 3 \\ +\ 5\ 0\ 6\ 6 \\ \hline \end{array}$$

9.
$$\begin{array}{r} 9\ 5\ 2 \\ 3\ 8\ 1 \\ +\ \ \ 6\ 3 \\ \hline \end{array}$$

Write the number of minutes shown by each clock.

10.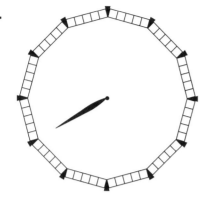

_____

11.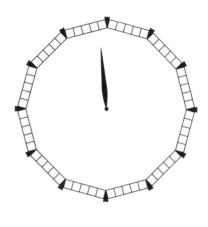

_____

12. Elaine has 35 thank-you notes to write. She wrote 9 notes yesterday and 9 more today. How many notes does she still have to write?

_____

Subtract, using regrouping when needed. Check by adding.

```
1.    6 8          2.    4 4
    - 1 9              - 3 6


3.    8 3          4.    7 2
    - 5 8              - 1 6


5.    2 5          6.    9 6
    -   9              - 4 7


7.    8 9          8.    9 4 6
    - 2 5              - 6 3 2


9.    7 5 1
    - 3 2 0
```

Add.

10.
```
  4 8 2 6
  1 4 9 3
+ 5 0 6 6
```

11.
```
  3 9 2 6
  3 2 3 8
+ 1 2 2 1
```

12.
```
  9 5 2
  3 8 1
  3 8 1
+   6 3
```

Write the number and say it.

13. 50,000 + 6,000 + 100 + 40 + 2 _____

14. Two hundred twenty-four thousand, six hundred fifty-one

_____

Round to the nearest thousand.

15. 3,016 → _____

16. 4,600 → _____

Write the number of minutes shown by each clock.

17.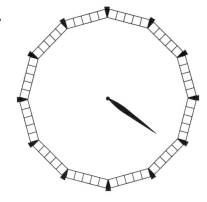

      _____

18.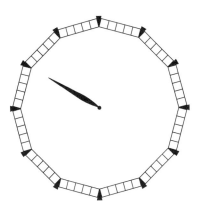

      _____

19. Last night 2,452 people came in cars to see the fireworks. The same night, 1,079 people walked to the fireworks. Another 958 people could see the fireworks from their homes. How many people viewed the fireworks in all?

_____

20. Earl earned $255 last winter shoveling snow. His mom gave him $125 for Christmas. Earl decided to send $140 to help others. How much money does he have left?

_____

Give the time with hours and minutes.

1.

_____

2.

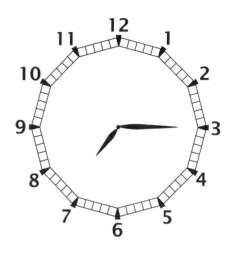

_____

3.

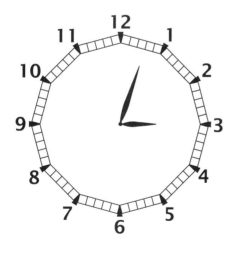

_____

4.

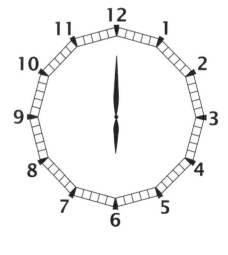

_____

Subtract and check.

5.
$$\begin{array}{r} 9\ 0 \\ -\ 1\ 3 \\ \hline \end{array}$$

6.
$$\begin{array}{r} 4\ 4 \\ -\ 3\ 5 \\ \hline \end{array}$$

7.
$$\begin{array}{r} 5\ 2 \\ -\ 2\ 9 \\ \hline \end{array}$$

8.
$$\begin{array}{r} 6\ 3\ 8 \\ -\ 1\ 1\ 7 \\ \hline \end{array}$$

9. Justina bought a magazine for $3.78 and a candy bar for $0.55. How much did Justina spend in all?

   _____

10. Nick has 3 dimes, 5 nickels, and 16 pennies. How much money does he have?

    _____

Subtract, using regrouping as needed. Check by adding.

1.　　2 0 0
　　－　3 8

2.　　3 6 7
　　－1 4 9

3.　　7 5 5
　　－2 8 6

4.　　8 8 0
　　－　9 4

5.　　4 8
　　－1 4

6.　　5 3
　　－4 2

Give the time with hours and minutes.

7.

————————

8.
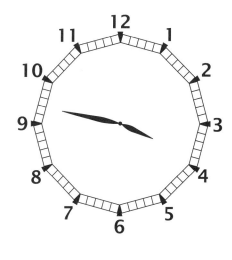

————————

Add.

9.   6 5 0
     + 1 6 3

10.   2 7 4
    + 8 9 5

11.   3 0 3 9
    + 5 2 1 1

12. Brad is reading a book with 313 pages. He has already read 194 pages. How many are left to read?

_____

13. Linda bought 10 feet of red ribbon, 12 feet of green ribbon, and 15 feet of gold ribbon to make holiday decorations. She has used 19 feet of ribbon so far. How many feet of ribbon does she have left?

_____

14. Steve flew 3,100 miles yesterday and 4,916 miles today. How many miles did he fly the last two days?

_____

Fill in the blanks.

1. The second month is _____.

2. The _____ month is April.

3. The _____ month is December.

Matching.

4. first   Friday

5. second   Tuesday

6. third   Sunday

7. fourth   Wednesday

8. fifth   Monday

9. sixth   Saturday

10. seventh   Thursday

How many days are in each month?

11. September _____

12. July _____

Tell how many.

13. ⵑⵑⵑ ⵑⵑⵑ ⵑⵑⵑ _____

14. ⵑⵑⵑ ⵑⵑⵑ || _____

Show the number with tally marks.

15. 18 _____

16. 25 _____

Add or subtract.

17.
```
  3 4 1
-   6 9
```

18.
```
  7 4 3
- 2 0 5
```

19.
```
  6 0 0
-   3 5
```

20.
```
  5 | 1 | 5 | 8
+ 1 | 2 | 2 | 2
```

Subtract. Check by adding.

1.
```
  7 8 1 0
-   6 8 1
```

2.
```
  5 1 2 3
- 4 8 8 2
```

3.
```
  2 1 7 5
- 1 5 0 4
```

4.
```
  4 9 3
-   5 8
```

5.
```
  8 2 4
- 3 6 7
```

6.
```
  4 1
- 3 2
```

Show the number with tally marks.

7. 10 _____

8. 32 _____

9. 28 _____

Give the time with hours and minutes.

10.

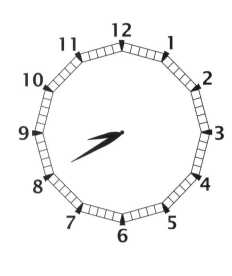

_____

11.

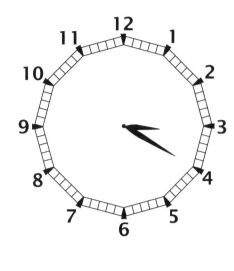

_____

12. _____ is the sixth day of the week.

13. February is the _____ month of the year.

14. What is the perimeter of a square that measures 43 inches on a side?

    _____

15. Abbie, Darcy, and Ben went out to pick strawberries. Abbie picked 5 quarts, Darcy picked 8 quarts, and Ben picked 13 quarts. Mom put 18 quarts in the freezer for next winter. How many quarts are left to eat now?

    _____

Add or subtract.

1. $5.67
    − 0.28

2. $3.19
    − 1.42

3.
    | 6 | 3 | 4 | 0 |
    − | 3 | 5 | 0 | 6 |

4.   175
    + 48

5. $4.29
    + 6.88

6.
    | 7 | 3 | 8 | 1 |
    | 2 | 4 | 7 | 9 |
    + | 5 | 6 | 3 | 0 |

Matching.

7. first ————————————— January

8. second   April

9. third   May

10. fourth   February

11. fifth   March

12. sixth   June

Skip count by two.

13. ____, ____, ____, ____, ____, ____, ____, 16, ____, ____

14. Use tally marks to show the number of days in a week.

    _____

15. The sandwich Deb wants to buy costs $3.50. She has a coupon for $1.75 off the price. What will Deb have to pay for the sandwich?

    _____

Subtract and check.

1. 6 3 , 4 5 2
   − 4 1 , 3 6 9

2. 7 5 , 8 3 0
   − 2 2 , 5 3 6

3. $1 4 . 9 8
   −   6 . 1 2

4. $5 9 . 2 0
   − 3 0 . 1 5

5. $9 1 . 4 2
   − 1 5 . 1 7

Fill in the oval with >, <, or =.

6. 26 $\bigcirc$ 62

7. 11 − 9 $\bigcirc$ 3 + 1

8. 4 + 5 $\bigcirc$ 5 + 4

Add

9.
```
  1|2|4
  3|5|6
  2|7|9
+ 4|5|1
```

10.
```
  7|2|9
  5|1|2
  2|8|3
+  |4|5
```

11.
```
  1|0|9
  4|6|0
  9|9|9
+  |3|1
```

12. _____ is the second day of the week.

13. February is the _____ month of the year.

14. Alan went to the store with $54.98 and came out with $21.15. How much money did he spend in the store?

    _____

15. At first, 19,345 tadpoles were swimming in the pond. Before they could turn into frogs, 18,400 of them were eaten. How many frogs hopped out of the pond that summer?

    _____

Read the gauge or thermometer. Write your answer on the line.

1.

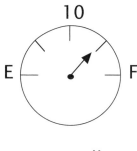

_____ gallons

2.

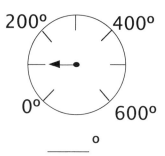

_____ °

3.

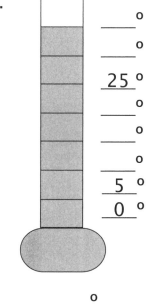

_____ °

4.

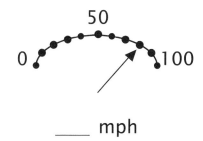

_____ mph

5.

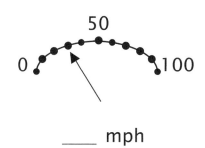

_____ mph

Add or subtract.

6.
$$\begin{array}{r} 5\,4\,0\,2\,1 \\ -\,1\,4\,0\,1\,5 \\ \hline \end{array}$$

7.
$$\begin{array}{r} 7\,2\,4\,3 \\ -\,3\,5\,6\,7 \\ \hline \end{array}$$

8.
$$\begin{array}{r} 9\,4\,6\,1 \\ +\,2\,7\,4\,3 \\ \hline \end{array}$$

Give the time with hours and minutes.

9.

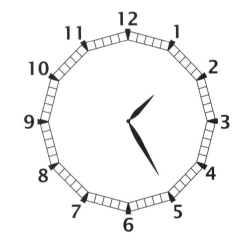

_____

10.

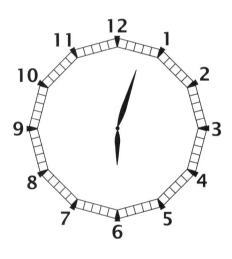

_____

11. George Washington was born in 1732 and died in 1799. How long did he live?

_____

12. Nicholas counted 47 cars on the way to the store. Use tally marks to show the number of cars he counted.

_____

Use the graph to answer the questions.

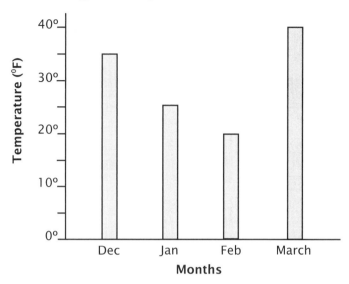

Average Temperature Per Month

1. Which month was the coldest? _____

2. Which month was the warmest? _____

3. What was the average temperature in January? _____

4. What was the average temperature in March? _____

Read the gauge or thermometer. Write your answer on the line.

5.

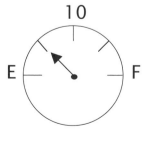

____ gallons

6.

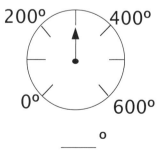

____ °

7.

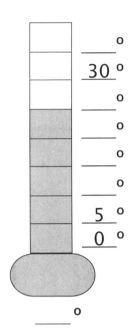

30 °

5 °

0 °

_____ °

8.

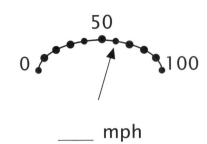

50

0   100

_____ mph

9.

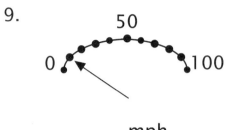

50

0   100

_____ mph

Add or subtract.

10.
```
  1 4 5 6 7
- 1 2 2 7 8
```

11.
```
  4 5 0 0
- 2 6 9 9
```

12.
```
  4 8 3 0
  9 4 7 1
+ 3 0 2 9
```

Subtract.

1.
```
    6 3 2
  - 4 4 9
```

2.
```
    7 5 0
  - 5 3 6
```

3.
```
    8 1 9
  - 2 6 3
```

4.
```
    5 2 0 8
  - 1 6 1 9
```

5.
```
    9 8 3 4 2
  - 6 8 4 5 3
```

6.
```
    $7 2.1 4
  -   3 4.2 1
```

Give the time with hours and minutes.

7.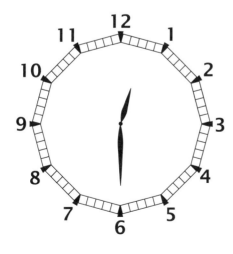

_____

8.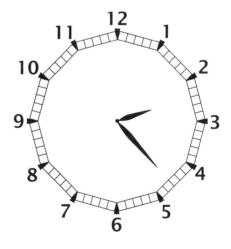

_____

Tell how many.

9. ‖‖‖ ‖‖‖ ‖‖‖ ‖‖‖ ‖‖‖ _____

10. ‖‖‖ |||| _____

Show the number with tally marks.

11. 19 _____

12. 31 _____

Fill in the blanks.

13. _____ is the fifth day of the week.

14. _____ is the third day of the week.

15. April is the _____ month of the year.

16. December is the _____ month of the year.

Use the graph to answer the questions.

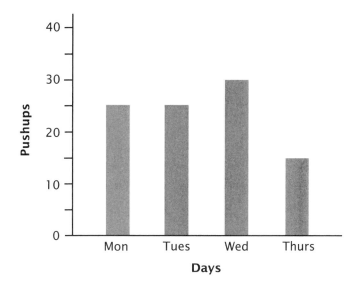

Pushups Jerry Did Per Day

17. On which day did Jerry do the most pushups?

    _____

18. How many pushups did he do on Thursday?

    _____

19. On which two days did he do the same number of pushups?

    _____ and _____

Read the gauge or thermometer. Write your answer on the line.

20.

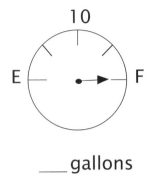

___ gallons

21.

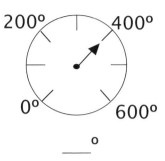

___ °

22.

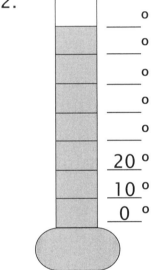

23.

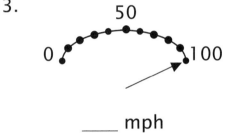

___ mph

24.

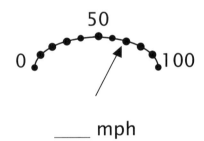

___ mph

Compare. Then fill in the oval with <, >, or =.

1. 7 + 7 $\bigcirc$ 15 – 8

2. 105 $\bigcirc$ 125

Round to the nearest ten.

3. 43 → _____

4. 68 → _____

Round to the nearest hundred.

5. 204 → _____

6. 561 → _____

Round to the nearest thousand.

7. 1,935 → _____

8. 4,187 → _____

Skip count and write the numbers.

9. ____, ____, 6, ____, ____, ____, ____, ____, ____, 20

10. ____, ____, ____, ____, 25, ____, ____, ____, ____, ____

11. ____, 20, ____, ____, ____, ____, ____, 80, ____, ____

Add.

12.   2 4
      + 4 6

13.   1 9 2
      + 3 5 9

14.   9 0 7
      + 1 6 8

15. $8.9 2
      +   2.4 9

16.   6 | 4 | 7 | 4
        7 | 6 | 1 | 0
      + 3 | 6 | 8 | 5

17.   9 | 6 | 8
        1 | 4 | 5
        2 | 0 | 3
      + | 7 | 5

Subtract.

18.   2 3
      − 1 7

19.   1 1 5
      −   9 8

20.   4 0 3
      − 2 1 5

21.   7 1 0
      − 3 4 6

22.   5 | 8 | 3 | 4
    – 1 | 0 | 5 | 7

23.   8 | 1 | 3 | 2 | 7
    – 4 | 5 | 1 | 8 | 9

Write the number and say it.

24.   Two hundred seventy-six thousand, five hundred ninety-one

_____

Give the time with hours and minutes.

25.

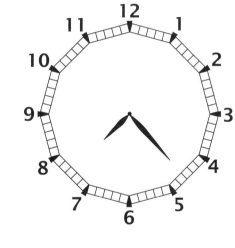

_____

Read the gauge.

26.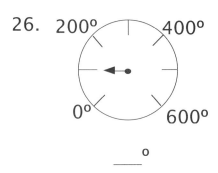

_____ °

27. Greg is four feet tall. How many inches tall is Greg?

    _____

28. Wendy is making a square pillow that measures 14 inches on a side. How many inches of fringe will she need to go around the edges of the pillow?

    _____

29 A rectangle has two sides that are six feet long and two sides that are eleven feet long. What is its perimeter?

    _____

30. A triangle has sides of nine inches, eight inches, and six inches. What is the perimeter?

    _____